NIST Special Publication 250-74

Absorbed Dose to Water Calibration of Ionization Chambers in a ^{60}Co Gamma-Ray Beam

Ronaldo Minniti

Jileen Shobe*

Stephen M. Seltzer

Heather Chen-Mayer

Steve R. Domen

Ionizing Radiation Division
Physics Laboratory
National Institute of Standards and Technology
Gaithersburg, MD 20899

Supersedes NIST Special Publication 250-40 (1990)

September 2006

---

* Retired from NIST in December 2002

Abstract

Absorbed-dose-to-water calibrations are important to the medical community to facilitate the accurate determination of doses delivered to tumors during external-beam cancer therapy. The first version of this document published in 1990 (NBS Special Publication 250-40) offered an absorbed-dose-to-water calibration service based on a graphite calorimeter as the primary standard instrument. However, the use of this calorimeter necessitated calculations to convert the measurement from graphite to water. In 1989, a water calorimeter was introduced at the National Institute of Standards and Technology (NIST) developed by Steve Domen, which was to replace the graphite calorimeter as the primary standard instrument. The calculations necessary for conversion factors were eliminated with this new technology. In 1995 NIST started to develop a new calibration service, based on the water calorimeter measurements, for disseminating the standard for absorbed dose to water. This document describes the calibration service offered by NIST for almost a decade and is based on the Domen water calorimeter standard.

Despite the fact that the service became available at NIST in the past decade, the medical-physics community did not take advantage of it during the first years and continued instead to have chambers calibrated in terms of the quantity exposure (in units of roentgen) to calibrate their ^{60}Co radiotherapy and high-energy x-ray producing electron accelerators. A protocol, commonly known as TG21, developed by the American Association of Physicists in Medicine (AAPM), involved many calculations to arrive at the quantity desired by the medical physicist in the practicing clinic, cGy/MU (centiGray/monitor unit). The AAPM developed a new protocol through Task Group 51 which involved absorbed-dose-to-water calibrations of ion chambers commonly used in the calibration of clinical radiotherapy photon and electron beams. The community now follows this protocol and makes use of the new NIST calibration service based on the water calorimeter standard.

NIST has developed and offers the absorbed-dose-to-water calibration service for ionization chambers based on a water calorimeter standard developed by Steve Domen at NIST. This document outlines the steps that have been taken to develop this service including a brief description of the Domen water calorimeter. The procedures that are involved in the calibration of an ionization chamber for this quantity are presented along with results from recent comparisons of the NIST with the Bureau International des Poids et Mesures (BIPM) in France and the National Research Council Canada (NRCC).

List of Figures

Figure 1	Vertical-beam ^{60}Co source.	5
Figure 2	Field size and beam uniformity at the level of the detector.	6
Figure 3	Schematic diagram of absorbed-dose-to-water calibration setup.	8
Figure 4	Schematic cross section of the Domen water calorimeter showing the essential features for measuring absorbed dose to water.	10
Figure 5	Schematic diagram showing the constructional details of the temperature probe.	11
Figure 6	Calibration data of thermistor resistances as a function of temperature.	13
Figure 7	Waterproofing assembly for chambers which are not inherently waterproof.	15
Figure 8	Water-phantom chamber-mounting apparatus.	15
Figure 9	Stabilization profile of the NE 2571 chamber.	17
Figure 10	Depth-dose curve for the NIST ^{60}Co source.	19
Figure 11	Absorbed-dose-to-water dependence on jaw settings.	20
Figure 12	Net Current as a function of chamber depth, normalized to the current at 5 cm depth. Source-to-surface distance (SSD) is maintained at 95 cm.	23
Figure 13	Net Current as a function of depth of water above chamber relative to the current with 5 cm water above the chamber. Source-to-detector distance (SDD) is held constant at a nominal 100 cm.	24
Figure 14	Dependence of measurements on height of chamber above tank bottom.	25

List of Tables

Table 1	Specifications of typical farmer-type ionization chambers used at NIST.	13
Table 2	Chamber characteristics.	16
Table 3	Uncertainty analysis for the primary-standard measurement of the absorbed-dose-to-water rate, $\dot{D}_w$. The relative uncertainties are expressed in %.	32
Table 4	Uncertainty analysis for the calibration of an ionization chamber in terms of absorbed dose to water. The relative uncertainties are expressed in %.	33
Table A.1	Results of the NIST-BIPM comparison	36
Table B.1	Results of the NIST-NRCC comparison	39

1. Scope

This document is intended to give a detailed report on the service offered by NIST for the calibration of ionization chambers in terms of the quantity absorbed dose to water. Ionization chambers are irradiated inside a water phantom using ^{60}Co gamma-ray beams. The NIST calibration code for this service is 46110 and appears listed in Chapter 8 under section C.1 of the NIST Calibration Services Users Guide SP 250.

2. Definitions and Acronyms

AAPM: American Association of Physicists in Medicine.

absorbed dose to water: The energy imparted by ionizing radiation per unit mass of water.

accreditation: A formal recognition that a laboratory is competent to carry out specific tests or calibration, or types of tests or calibrations.

ADCL: Accredited Dosimetry Calibration Laboratory

air kerma: Air kerma, K, is the quotient of dE_{tr} by dm, where dE_{tr} is the sum of the initial kinetic energies of all electrons liberated by photons in a volume element of air and dm is the mass of air in that volume element. Then

$$K = \frac{dE_{tr}}{dm} \qquad (1)$$

The SI unit of air kerma is the gray (Gy), which equals one joule per kilogram; the older unit of air kerma is the rad, which equals 0.01 Gy.

exposure: Exposure is defined as the total charge per unit mass liberated in air by a photon beam and is represented by the equation:

$$X = \frac{dQ}{dm} \qquad (2)$$

where dQ is the sum of the electrical charges of all the ions of one sign (negative or positive) produced in air when all the electrons liberated by photons in a volume element of air whose mass is dm are completely stopped in air. The SI unit of exposure is the coulomb per kilogram (C/kg); the special unit of exposure, the roentgen (R), is equal to exactly 2.58×10^{-4} C/kg. The ionization arising from the absorption of bremsstrahlung emitted by the secondary electrons is not included in dQ. Except for this small difference, significant only at high energies, the exposure as defined above is the ionization equivalent of air kerma. The relationship between air kerma (in Gy) and exposure (in

R) can be expressed as a simple equation:

$$K = X \ (2.58 \times 10^{-4} \frac{C}{kg})\left(\frac{W}{e}\right)\left(\frac{1}{1-g}\right) \tag{3}$$

where W/e is the mean energy per unit charge expended in dry air by electrons, and g is the mean fraction of the initial kinetic energy of secondary electrons liberated by photons that are lost through radiative processes in air. The currently accepted g values for ^{60}Co, ^{137}Cs and x-ray beams are 0.0032, 0.0016 and 0.0000 respectively[1]. The current value used by the NIST for W/e is 33.97 J/C, which is the value currently adopted by the international measurement community[2].

BIPM: Bureau International des Poids et Mesures.

calibration: The set of operations that establish, under specific conditions, the relationship between values indicated by a measuring instrument or measuring system, or values represented by a material, and the corresponding values realized by standards.

calibration coefficient, $N_{D,w}$: The *absorbed-dose-to-water calibration coefficient*, $N_{D,w}$, for an ionization chamber is defined as the quotient of the value of the absorbed dose to water delivered to the chamber and the electrical charge generated by the radiation in the ionization chamber. Equivalently it can be defined as the quotient of the absorbed-dose-to-water rate delivered to the chamber and the ionization current generated by the radiation in the ionization chamber. The units of the calibration coefficient are given in Gy/C.

calibration factor: The result of a calibration when the ratio of values realized by the test instrument and the standard is dimensionless.

interlaboratory comparison: A program to provide organization, performance and evaluation of calibrations or tests on the same or similar items or materials by two or more laboratories in accordance with predetermined conditions.

ionization chamber: A solid envelope surrounding a gas- (usually air-) filled cavity in which an electric field is established to collect the ions formed by the radiation.

measurement: The set of operations having the object of determining the value of a quantity.

measurement system: One or more measurement devices and any other necessary system elements interconnected to perform a complete measurement from the first operation to the result.

NIST: National Institute of Standards and Technology

NRCC: National Research Council - Canada.

proficiency testing: A test conducted by NIST to demonstrate that laboratories participating in the test are able to transfer a measurement of absorbed dose to water that is traceable to NIST.

temperature-pressure correction factor k_{TP}: The correction factor k_{TP} is computed from the following expression: $k_{TP} = (273.15 + T)/(295.15H)$ where T is the temperature in degrees celsius, and H is the pressure expressed as a fraction of a standard atmosphere. (1 standard atmosphere = 101.325 kilopascals = 1013.25 millibars = 760 millimeters of mercury).

uncertainty of measurement: An estimate characterizing the range of values, within an approximate level of confidence, in which the true value of a measurand lies.

verification: Conformation by examination and provision of evidence that specified system requirements have been met. Verification includes all sub-system tests.

3. Standards and Facilities

3.1 NIST absorbed-dose-to-water standard

There have been adjustments to the NIST standard for absorbed dose to water from ^{60}Co radiation during the last few years, associated mainly with the development of the water calorimeter. This section documents these small changes and documents the present standard that is described more fully in Section 4.

Before 1991. Prior to 1991, the NIST standard for absorbed dose to water from ^{60}Co radiation was based on measurements performed with a graphite calorimeter[3,4].

1991. In early 1990, Domen began measurements with his newly developed, N_2-saturated, sealed-water calorimeter[5]. Because this system was new, the measurements were confirmed by comparison with those made with a graphite calorimeter and with a graphite-water calorimeter. Based on the very good agreement among these results, a comment in Domen's paper suggested the use of an average of the measurement results from these three different NIST calorimetry systems. Such a so-called "blended" standard for NIST was used in one bilateral comparison of standards with the NRCC in 1991.

1992. Subsequent to the 1991 NRCC comparison, Domen investigated the H_2-saturated, sealed-water calorimeter and found it superior to the N_2-saturated system[5].

1994. In 1994, the result of the 1991 NIST/NRCC comparison[6] was related to the BIPM standard (and thus to other national standards at that time) through the result of a NRCC/BIPM comparison. However, Boutillon et al.[7] employed a blended standard for NIST consisting of the average of the results from the H_2-saturated, sealed-water calorimeter, the graphite-water calorimeter, and the graphite calorimeter, rather than the average that included the N_2-saturated system as used in 1991. The absorbed dose rate in water from the NIST ^{60}Co source intrinsic to this 1994 blended standard was 0.99956 of that of the 1991 blended standard. It is unclear if this slight change of standards was taken into account, but the 0.04% difference would seem to have negligible significance.

1995. In 1995, NIST began preparation to disseminate the absorbed-dose-to-water determinations of Domen. The perspective had by then changed. Rather than using graphite calorimetry systems to confirm water-calorimetry results, it became increasingly accepted that the water-calorimetry results were of the highest metrological value and, being a more direct determination of the quantity of interest, could be used to confirm results of graphite calorimetry. It was thus decided to use only a result obtained with the sealed-water calorimeter. However, the result chosen was that from the N_2-saturated, sealed-water calorimeter. This implied an absorbed dose rate in water from the NIST ^{60}Co source that was 0.99983 of that using the 1994 blended standard.

1998. From a review of procedures in consultation with Steve Domen, it was concluded that Domens' results with the H_2-saturated, sealed-water calorimetry system constitute the most direct and technically superior of the NIST measurements of the absorbed dose in water, and thus is most suitable for the standard. This standard then implied an absorbed dose rate in water from the NIST ^{60}Co source that was 0.99868 of that using the N_2-saturated result. This 0.13% difference in the measurement results from the N_2 and the H_2 systems is not significant (compared to the uncertainty of the measurement discussed later), nor for that matter is the 0.15% difference between the 1994 blended standard and the current H_2-saturated, sealed-water standard. But consistency requires that a slight adjustment in transfer standards be made. Thus, NIST absorbed-dose-to-water calibrations reported prior to 25 June 1999 (and after 1995) should be multiplied by the factor 0.99868.

3.2 ^{60}Co source

The ^{60}Co source is housed in an Atomic Energy of Canada, Ltd. (AECL)[a] Theratron F teletherapy head that is mounted to the ceiling. This source configuration produces a vertical gamma-ray beam as shown schematically in Fig. 1. The Theratron head contains adjustable jaws for collimation of the beam. For calibration work requiring a known absorbed-dose-to-water

---

[a]Certain commercial equipment, instruments, or materials are identified in this report to foster understanding. Such identification does not imply recommendation or endorsement by the National Institute of Standards and Technology, nor does it imply that the materials or equipment identified are necessarily the best available for the purpose.

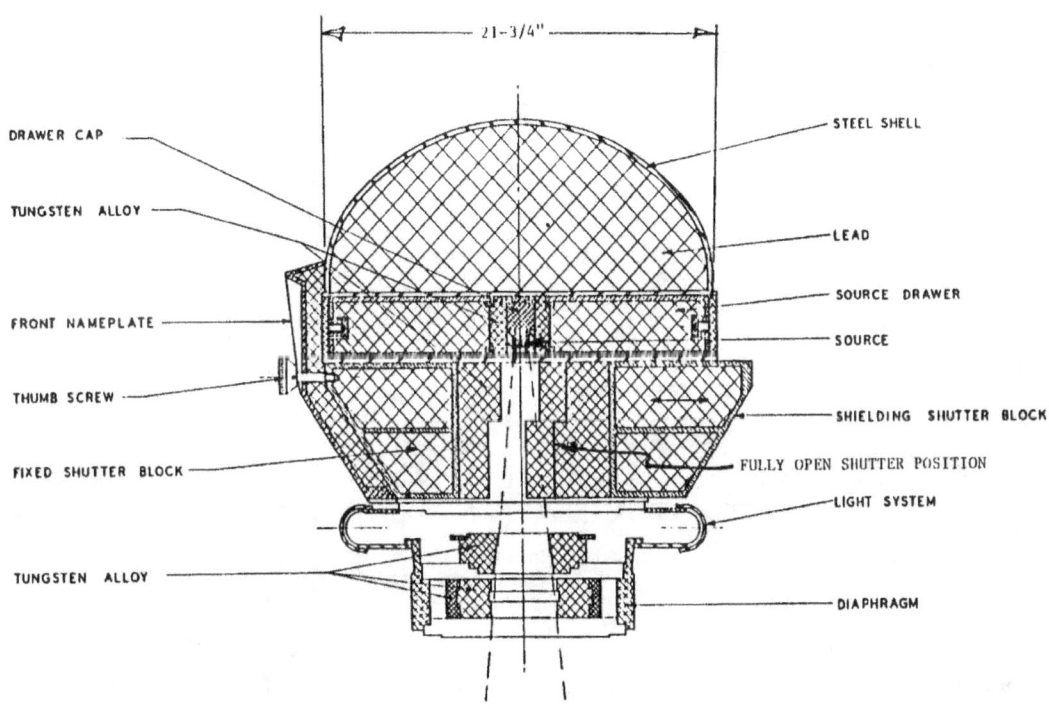

Figure 1. Vertical-beam ^{60}Co source.

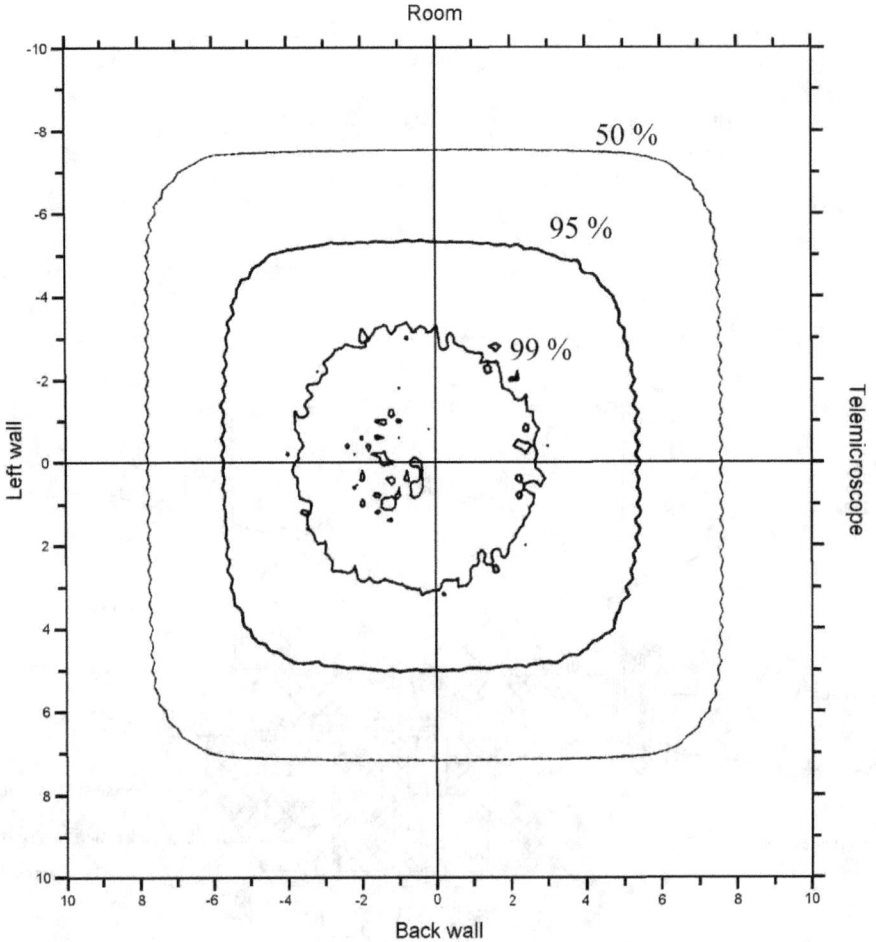

Figure 2. Field size and beam uniformity at the level of the detector with the jaws set with an aluminum block. This view looks up toward the source. The cross hairs represent the geometrical center of the beam. The field size was determined by use of a CMS, Inc. scanning water phantom. Dimensions are in centimeters.

radiation field, measurements are conducted with a square collimator opening of 50.8 mm by 50.8 mm defined by closing two pair of jaws onto a block of those dimensions. Using a Computerized Medical Systems (CMS), Inc. scanning water phantom, it was determined that this collimator setting defines a field size of 15.4 cm x 15.4 cm (50 % intensity line) at the level of the detector, Fig. 2.

3.3 Calibration facilities

NIST maintains a vertical 444 TBq (nominal 12 kCi Nov, 1989) ^{60}Co source in the low intensity vertical beam facility located where the absorbed-dose-to-water rate is known as a function of field size, chamber depth in water, and distance from the source. The source is located approximately 2.6 m above the floor. Directly under the source is a pit covered by an aluminum-magnesium alloy low-scatter grate. The pit is approximately 2 m deep and 1.5 m in diameter to help reduce scatter from the well-collimated beam. A laser, located in the base of the pit, is used to define the central axis of the beam and for alignment of the test chamber.

The control room contains all peripheral equipment such as electrometers, temperature and pressure readout units, and the computer that controls the calibration system. Customer chambers are kept in this room from the time they are received until they are shipped back to their owners.

NIST is currently set up to calibrate farmer-type-chamber models: Exradin A12, NE 2571, and PTW N23333 without customer electrometers. Other makes and models of chambers can be calibrated as long as the customer provides a suitable waterproofing cap for the chamber, if not inherently waterproof, and the chamber/cap combination must fit within the NIST water phantom, see Section 5.

3.4 The calibration service

The absorbed dose rate was determined by use of the water calorimeter placed in the beam with the following source field parameters: $s = 5.08$ cm, $x = 5.0$ cm, $z =$ nominal 100 cm, and $f = 15.4$ cm as shown in Fig. 3. As described in Section 3.1, the value of the measured absorbed dose to water rate was subjected to several corrections up to 1999. The value of the initial measurement, made on January 11, 1990, including all corrections applied until June 25 1999, is $\dot{D}(0) = 1.812$ Gy/min. This value has been used since 1999 to determine the absorbed-dose-to-water rate, $\dot{D}(t)$, for any given day of the year by using the following expression,

$$\dot{D}(t) = \dot{D}(0) \cdot exp\left(-3.60056 \times 10^{-4} \cdot t / days\right), \qquad (2)$$

where t is the number of days between the reference date and any other given date of the year. The absorbed-dose-to-water rate measurement done with the water calorimeter is then transferred to ionization chambers placed in a water phantom with the same source field parameters (note that both dimensions z and x from Fig. 3 are measured with respect to the physical center of the collecting volume for cylindrical or spherical detectors). When the ionization chamber to be calibrated is placed at the same s, x, z, f as above, it can be calibrated in

terms of absorbed dose to

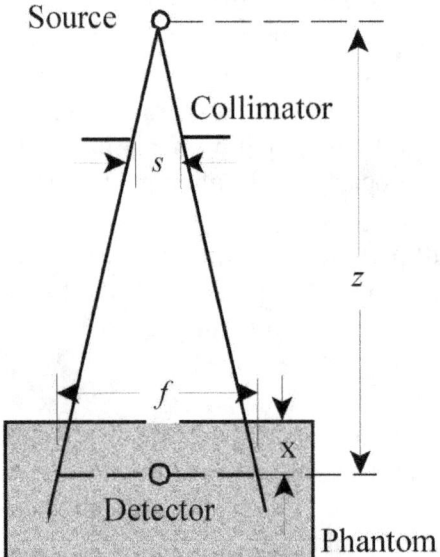

Figure 3. Schematic diagram of absorbed-dose-to-water calibration setup showing collimator size (s), detector depth in phantom (x), source-to-detector distance (SDD) (z), and field size at the detector (f).

water per unit charge (Gy/C). It can then be used in other ^{60}Co beams with similar geometries to determine absorbed-dose-to-water rates.

The NIST ionization chambers are used in the quality-assurance program to verify the absorbed-dose-to-water rate by recording over time the value of the calibration coefficients for several chambers. The calibration coefficient, $N_{D,w}$, can be determined for a given ionization chamber as,

$$N_{D,w} = \frac{\dot{D}}{I}, \tag{3}$$

where $\dot{D}$ is the rate of absorbed dose to water (in units of Gy/s) to which the ionization chamber was exposed, and I is the measured value of the ionization current (in units of C/s) of the ionization chamber.

The measured value of I includes corrections to account for deviations of the ambient temperature and pressure from standard conditions and a correction for electronic-recombination effects. With the assumption that the chamber is open to the atmosphere, the measurements are normalized to a pressure of one standard atmosphere (101.325 kPa) and a temperature of 295.15 K (22 °C). Use of the chamber at other pressures and temperatures requires normalization of the ion currents to these reference conditions by applying the correction factor k_{TP} introduced in the definition section (Section 2). A correction factor k_{sat} to account for ionization loss due to electronic recombination can also be applied to the measured ionization current. However, corrections for electronic recombination are not applied to chambers sent to NIST for calibration. Customers are informed that the calibration coefficient reported can be transferred to other beam facilities provided that the dose rates are similar to those used at NIST during the calibration of the instrument. If absorbed-dose-to-water rates are significantly different from those used for the calibration, a detailed study must be performed by the user of the instrument to correct for recombination loss. NIST does provide, however, the ratio of currents at full and half collection potential. The value of this ratio can be used to determine a value of k_{sat} using the two-voltage method[8]

4. The Water Calorimeter

The water calorimeter and its use as a primary standard has been described in the Journal of Research of the National Institute of Standards and Technology in an article entitled "A Sealed Water Calorimeter for Measuring Absorbed Dose"[5]. A summary of the contents of that article is described in this section. The reader can refer to the original article for more details.

Figure 4 shows a schematic cross section of the calorimeter and its general features. The phantom is a 30 cm x 30 cm x 30 cm acrylic container filled with distilled water. The electrical resistivity of the water is about 0.4 MΩ·cm. A thermistor is mounted to measure the temperature of this water with a resolution of 0.01 °C.

A sealed, thin-wall, cylindrical glass container, 110 mm long x 33 mm diameter, is mounted on supports within the phantom. The glass container serves to seal in the high-purity water and acts as a convective barrier. This water is cleaner than normally distilled water, having been prepared in a system consisting of a filter, deionizer, and an organic absorber. It has an electrical resistivity of 20 MΩ·cm (at 20 °C). Once prepared, the water is withdrawn and handled in glassware that was cleaned and placed overnight in a furnace at 450 °C. The water is then saturated with high-purity hydrogen or nitrogen gas prior to being sealed in the container.

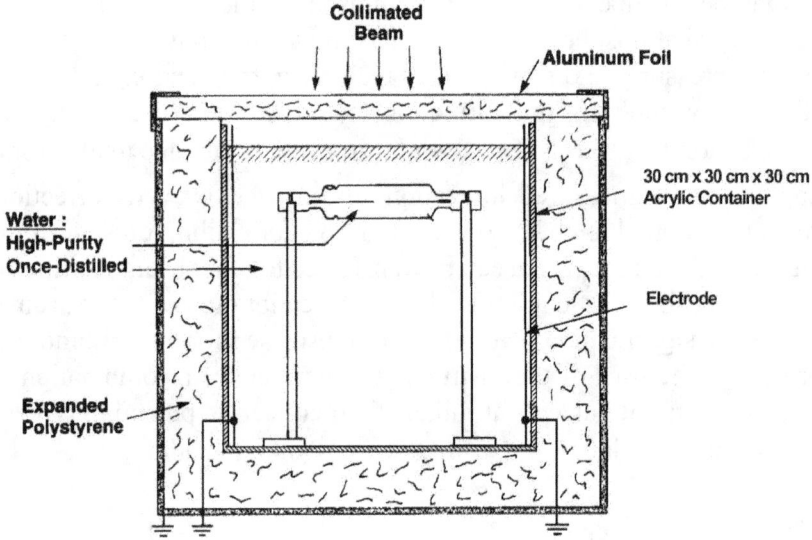

Figure 4. Schematic cross section of the Domen water calorimeter showing the essential features for measuring absorbed dose to water.

Two temperature probes are mounted within the glass container along the central axis with the sensor ends close to the center. Each probe consists of a thermistor that is enclosed within and near the end of a thin glass capillary. Figure 5 shows a schematic diagram giving constructional details of the sensor end of the probes. The distance between the thermistors can be varied, but was set at 9 mm for this work. The phantom is then filled with the once distilled water until the thermistors are at a depth of 5 cm from the surface. A temperature rise is produced by the square-collimated ^{60}Co beam.

The temperature rise is measured with the two calibrated thermistors in opposite arms of a Wheatstone bridge to double the output signal. For negligible heat defects and changes in thermistor power, the absorbed dose D is:

$$D = \frac{1}{2} \cdot (\Delta R/R) \cdot \left|\overline{S}^{-1}\right| \cdot c, \qquad (4)$$

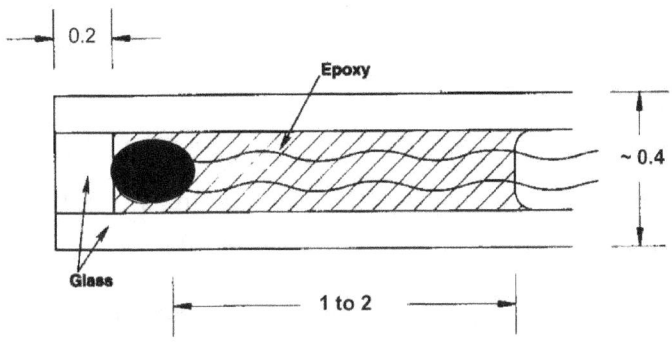

Figure 5. Schematic diagram showing the constructional details of the temperature probe consisting of an embedded thermistor near the end of a long thin capillary. Dimensions are in millimeters.

where the factor ½ is the result of using two thermistors to measure the temperature rise, $\Delta R/R$ is the measured fractional change in the Wheatstone bridge balancing resistor, $\left|\overline{S}^{-1}\right|$ is the absolute value of the reciprocal of the mean sensitivity of the thermistors determined from the calibration data, and c is the specific heat capacity of water at the calorimeter operating temperature. The product $\frac{1}{2} \cdot (\Delta R/R) \cdot \left|\overline{S}^{-1}\right|$ is the mean temperature rise.

The sensitivity of a thermistor, S, is defined as $(1/r)(dr/dT)$. S can therefore be expressed as,

$$S = \left|\beta/T^2\right|, \tag{5}$$

where β is the "material constant" of the thermistor. S is therefore determined from the value of β that is obtained by using the well known empirical expression:

$$r = r_o\, e^{\beta(1/T - 1/T_o)}, \tag{6}$$

where r and r_o are the values of the thermistor resistance at a given temperature T and T_o respectively (the unit of temperature is expressed in Kelvin).

Equation 6 can be reduced to the linear form,

$$y = \beta \cdot X + \theta, \qquad (7)$$

where $y = \ln r$, $x = 1/T$, and $\theta = \ln r_o - \beta/T_o$, a constant. Least-squares fits of the data are applied to Eq. 7 to determine β and ultimately the value of S.

Calibration of a thermistor consists of determining S, which is its fractional change in resistance per degree change in temperature. The temperature was measured with a calibrated mercury thermometer (0.01 °C per division) and with a quartz thermometer. The thermistors used had a resistance of about 3.3 kΩ at 22 °C with a negative coefficient of resistance (S) of about 3.7 %/K. The temperature probes were removed from the glass container and placed in the once-distilled water so that they would rapidly change with the water temperature, which was varied at intervals of 1 °C from 15° C to 29 °C by using ice and inmersion heaters. The temperature of the water was raised with four immersion heaters (total 100 W), and the water was circulated to attain uniform temperature. Then the water was allowed to become stagnant before measuring the thermistor resistances. Their resistances as a function of temperature are shown in Fig. 6. The bridge was balanced at each temperature. The thermistor resistances (r_1 and r_2) plus the external lead resistances, which were about 0.6 percent of r_1 or r_2, could be determined from the two known resistances on the bridge and the four measured potentials across the bridge arms as indicated in Fig. 4 of reference 5. The thermistors, however, were calibrated one at a time (for the sake of safety in handling), which required replacing a thermistor with a known resistance. This then gave two methods of determining a thermistor resistance from (1) two measured potentials and a known resistance, and (2) three known resistances. The electrical power in the thermistor varied from about 4 uW to 5 uW during calibration. The small rises in temperature of the thermistors were determined from measurements and added to the measured water temperature in the phantom.

A second generation Domen water calorimeter is currently under development at NIST. Preliminary results on this development have been presented recently at an international conference[9]

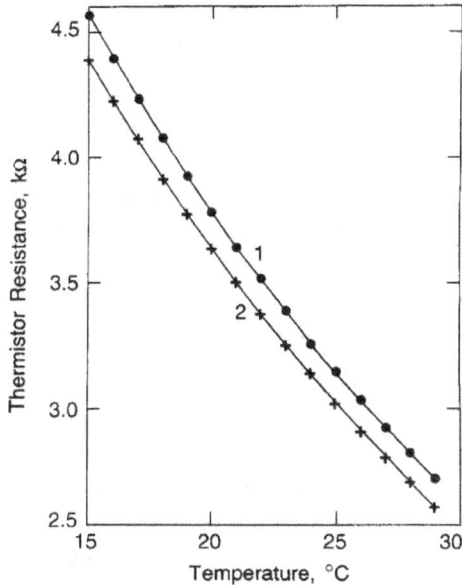

Figure 6 Calibration data of thermistor resistances as a function of temperature. The curves are marked for thermistor #1 and thermistor #2.

5. Secondary (transfer) standard ionization chambers

Farmer-type ionization chambers are used at NIST as secondary standards for consistency checks and as transfer standards. These types of chambers are most commonly used in radiotherapy clinics in the United States. Table 1 shows technical manufacturers' specifications for two typical farmer-type chambers.

Table 1. Specifications of typical farmer-type ionization chambers used at NIST.

Chamber	Nominal diameter (mm)	Nominal thimble length (mm)	Nominal body diameter (mm)	Thimble material	Nominal Collecting volume	Operating potential
Exradin A12	7.031	25.738	9.822	C552 plastic	0.65 cm^3	-300 V
NE 2571	6.977	25.717	8.613	graphite	0.60 cm^3	-300 V

The Exradin chambers are inherently waterproof while the NE chambers are not. In order to perform measurements with the NE chambers in water a two step assembly is required. First, a 1 mm thick acrylic sleeve is threaded onto the chamber stem in the same manner as the buildup caps. Next, latex sheaths are fastened with rubber bands to the back portion of the acrylic sleeve in such a manner that no portion of the sheath covers the collecting volume of the chamber. The purpose of the acrylic sleeve is to protect the chamber from the water while the sheath serves to prevent water from coming into contact with the cables. Figure 7 shows how the waterproofing is assembled for chambers that are not inherently waterproof. The flexible sheath is needed behind the acrylic sleeve due to the mounting configuration at NIST, which is shown in Fig. 8. Ross and Shortt[10] have shown that rubber-like waterproofing sheaths can cause a significant difference in the response of the chamber. Their work was repeated at NIST, as explained in the next section, with similar results.

Table 2 shows the reproducibility of both the system and the measurement setup. For the system reproducibility test, many contiguous calibrations of the chambers were performed by starting and stopping the data-acquisition program. For the setup reproducibility test, the water tank and chamber were reset in the beam for each calibration as if a new calibration were being initiated.

The positive and negative potential-response ratios can also be seen in Table 2. It is recognized that the response of the chamber can be different depending on the polarity of the potential applied[11,12,13], and that the correct ionization should be represented as:

$$M = \frac{(Q^+ - Q^-)}{2}, \qquad (8)$$

where Q^+ and Q^- represent the charge collected with positive and negative potentials respectively (I^+ and I^- are the corresponding ionization current values).

It is interesting to note that differences in manufacturing seem to be a prime contributor to the ratio of Q^+ and Q^-. While NIST operates only a limited number of chambers, calibrations of customer chambers show that the charge ratio for a given make and model is consistent. One criterion used to determine if chambers are operating normally is if their charge ratio corresponds to that common for their make. The NE 2571 ion chambers, for example, exhibit different polarity effects in water than they do in air as shown in Table 2.

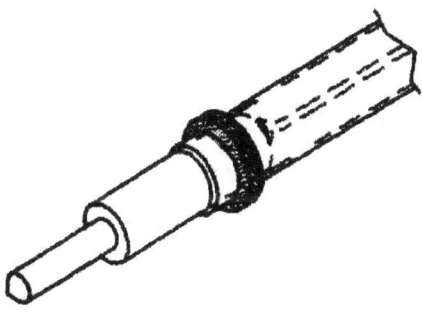

Figure 7. Waterproofing assembly for chambers which are not inherently waterproof. Drawing courtesy of Med-Tec.

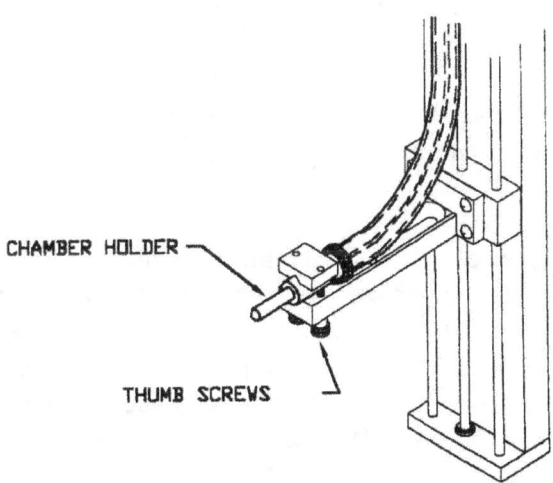

Figure 8. Water-phantom chamber-mounting apparatus. Drawing courtesy of Med-Tec.

Table 2. Chamber characteristics. Reproducibility measurements were made using a potential difference of 300 V.

Chamber	System Reproducibility (%)	Setup Reproducibility (%)	I^+/I^- air	I^+/I^- water
Exradin A12	0.01	0.05	1.0006	1.0006
NE 2571	0.01	0.03	1.0030	1.0020

During international comparisons, NIST obtained different polarity effects for the NE chambers for absorbed-dose-to-water measurements. NIST repeatedly observed an average 0.34 % polarity effect whereas other laboratories reported a polarity effect on the order of 0.15 %. At first, different field sizes at NIST were suspected. In the medical field, and hence most calibration laboratories, a field size of 10 cm x 10 cm at the detector is used. NIST's field size is 15.4 cm x 15.4 cm. The results of an investigation showed no field-size dependence as reported in the next section.

NE chambers are not fully guarded. Because of this, the stabilization time was investigated and the following test was developed: (a) a chamber was set up, the voltage (negative) was turned on, and the calibration was started; (b) the chamber set up was repeated (to account for evaporated water), the polarity switched to positive, and the calibrations repeated; and (c) after a repeat of the set up, the first conditions were repeated. The calibrations were made at three minute intervals over a two hour period for each condition. The entire test was repeated three times and each showed similar results.

As seen in Fig. 9, the first set of calibrations (a) were consistent for the entire calibration period with a standard deviation of 0.02 % over the two hours. The second set (b), however, shows the typical "settling" behavior observed with these chambers before coming to stability after more than an hour of measurements. Returning to negative potential (c), the chamber again exhibited a "settling" period of less than an hour. The final measurements of both (b) and (c) are on the order of 0.40 % higher than the initial measurements. The final negative-potential calibrations

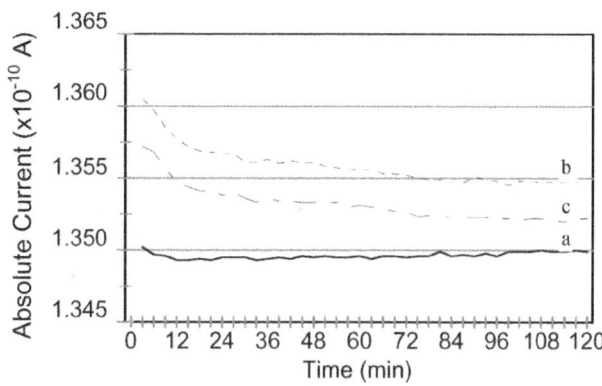

Figure 9 Stabilization profile of the NE 2571 chamber. Measurements were made every three minutes for two hours. (a) First set of measurements with polarity at -300 V. (b) Polarity switched to +300 V. (c) Final measurements at -300 V.

(c) were 0.15 % different than the initial measurements (a). The ratio of the final currents measured for (b) and (a) is 0.33 %, consistent with NIST's historical average reported in the table above. The ratio (b) to (c) is 0.18 %, consistent with that observed at other national laboratories. The observed effects here described could have contributed to the differences mentioned above observed during past comparisons.

The Farmer-type chambers discussed here are all open to the atmosphere. However, none of the chambers exhibit ideal atmospheric communication under rapidly changing conditions. The Exradin chambers have vent tubes that run from the chamber for the full length of cable, approximately 200 cm. As these chambers are inherently waterproof, the length of the vent tubes was probably seen as necessary to clear any possible situation in which the cable might lie in the water. Due to their small diameter, air exchange is not optimal, and they do not track pressure changes as quickly as desirable. Therefore, calibrations for Exradin chambers should not be made on days when there are rapid changes in pressure. The vent holes on the NE chambers are covered by the waterproofing caps. While this could pose a problem under rapidly changing atmospheric conditions, they still respond in a reasonable fashion.

The Exradin chambers are guarded and come to equilibrium almost immediately. Also, they do not show the polarity effect variations that the NE chambers do. Once equilibrium is reached, however, both the Exradin and the NE chambers are very suitable for calibration work.

An observation worth mentioning is in regard to the NE chambers: One of the NIST NE chambers began to show some erratic behaviour during one of the above mentioned calibrations. Taking the sleeve off, it was found that water had seeped into the waterproofing sleeve and the chamber was completely wet. Left open to the air to dry for approximately one week, the chamber response returned to normal.

6. Characterization of absorbed-dose calibration parameters

The NIST ^{60}Co source was calibrated with a water calorimeter, as described earlier, at a nominal source-to-detector distance of $z = 1.00$ m. A square metal jig with sides of 50.8 mm was fitted into the collimator opening, and the jaws were then closed onto it to set the beam size. Absorbed-dose-to-water measurements with the water calorimeter were made at only this one distance and collimator setting.

Since initial calibration, the source rate is determined through decay corrections using a half life of (1925.3 ± 0.5) days[14]. A NIST secondary-standard ionization chamber has tracked the source decay in terms of air kerma since the source was originally purchased in 1989. Recent success in air-kerma and absorbed-dose-to-water comparisons[15,16,17] show that the source decay is tracking well.

Beam size, uniformity, and percentage depth dose were determined by use of the CMS, Inc. scanning water phantom. In the first version of this document, SP250-40, Pruitt had calculated the field size at the detector to be 14.5 cm x 14.5 cm. The field size determined by the scanning water phantom is 15.4 cm x 15.4 cm at the 50 % limits. A scan of the field showing both beam size and uniformity can be seen in Fig. 2. The extreme values are within 5 % of each other over a field size of 12 cm x 12 cm and within 1 % over a field size of 4 cm x 4 cm. The percentage depth-dose can be seen in Fig. 10 and appears typical of that for a ^{60}Co source.

Several tests were performed to check the sensitivity of the ion chamber readings to certain measurement parameters such as field size, source-to-detector distance (SDD), surface-to-detector distance, chamber position relative to tank bottom, setup reproducibility, electrical leakage, water-proof materials, water purity and polarity effects. In the sections that follow a brief summary on each of these tests is described.

6.1 Field-size dependence

In this section, a test to study the dependence of the chamber reading (ionization current) with field size was determined. The field size is defined by two pair of collimator jaws located on the therapy unit. The two pair of jaws are perpendicular to each other. Each pair of jaws is set manually by adjusting the corresponding collimator dial. The readings on the dials are dimensionless and can be varied between 0 (jaws completely closed) and 20 (jaws completely open). The dial setting on both pair of jaws has to be set to the same value to obtain a square

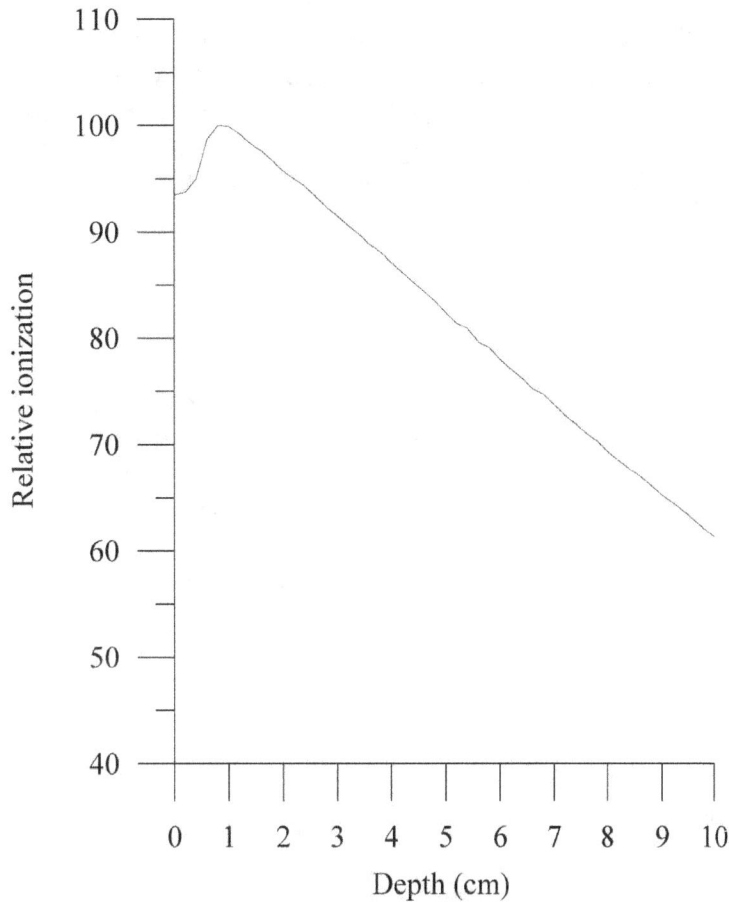

Figure 10. Depth-dose curve for the NIST ^{60}Co source determined by the CMS, Inc. scanning water phantom.

field. For chamber calibrations made in terms of absorbed dose to water a square field is used and the dial setting for both pair of jaws is 10.75 (this value has been used for setting the field size since the year 2001). Throughout the text we use the notation 10.75 x 10.75 to refer to a dial setting on both pair of jaws of 10.75. This collimator-dial setting defines approximately a square field size of 15.4 cm x 15.4 cm at a source-to-detector distance of 1 m as shown in Fig. 2.

For the field-size test the collimator-dial settings were varied from 10 x 10 to 11.5 x 11.5 in increments of 0.25. Field sizes (50 % contour at the level of the detector) associated with these settings vary from 14 cm x 14 cm to 16.1 cm x 16.1 cm. In Fig. 11 the readings at the various jaw settings are normalized to the reading obtained when the dials are set to 10.75 x 10.75. It was found, but is not shown, that the readings vary depending on whether the collimators are opened 'up' to the settings or closed 'down' to the settings. The procedure followed is to close down the jaws to the dial settings. From the figure one can infer that the reproducibility in the field size on a daily basis for chamber calibrations can be considered to be within 0.05 %.

6.2   Problems with field size[18]

In November 2000, it was discovered that there was a potential problem with the NIST absorbed-dose-to-water calibration service. For a chamber that had recently been returned from an international comparison, the calibration coefficient was significantly different than the historical calibration coefficient that had been verified prior to the comparison. Upon careful examination, it was found that the problem was due to differences in the radiation field size.

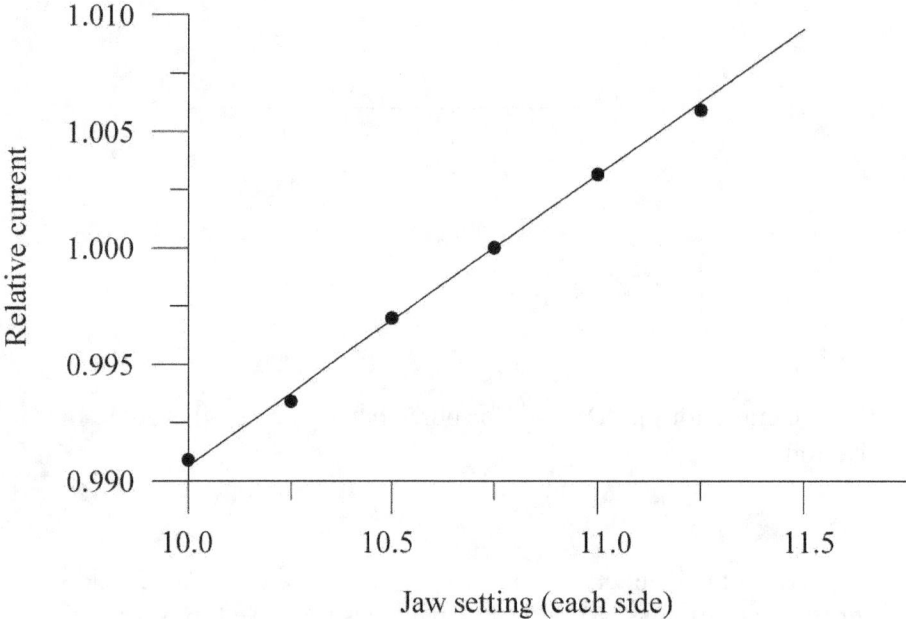

Figure 11. Absorbed-dose-to-water dependence on jaw settings. Each pair of jaws are set to the same dial setting (see text). The current response at each setting is shown relative to the that using the 10.75 x 10.75 dial setting.

NIST had used a square metal jig with sides of 50.8 mm to set the jaws on its ^{60}Co Theratron F

NIST had used a square metal jig with sides of 50.8 mm to set the jaws on its ^{60}Co Theratron F teletherapy unit. The jig was inserted into the opening, and the jaws were then closed around it to set the jaw openings and hence the field size. The unit has a set of four knobs to control so-called "beam trimmers", designed to cut individual corners of the field in therapy applications. The two knobs used to manually set the jaw openings are located between three of these trimmer knobs. It appears that two of the trimmer knobs, instead of the jaw knobs, had been accidentally turned, with the effect that the jig was held in place partially by the trimmers. This had the effect of changing the field size set by the jig. When the trimmers were subsequently backed out, the calibration coefficient was again significantly different than the historical value, but now in the opposite direction. The question then became, were the trimmers out of the way when water-calorimeter measurements were made to establish the absorbed-dose-to-water rate for the NIST source back in 1990?

After testing, it was determined that for the past three to four years before the end of 2000 during which the NIST absorbed-dose-to-water calibration service had been offered, and during which the historical calibration coefficients of the secondary standards were determined, there was at least one trimmer set slightly into the beam. The NIST absorbed-dose-to-water calibration coefficients had always shown a larger variation in reproducibility than those for air-kerma, and this was due to the fact that the jig has the potential of fitting in the jaws in a number of ways when a trimmer is in the way. It should be noted that this did not affect NIST air-kerma calibrations, as the jig is not used to set the jaws in that case and therefore, the trimmers would only cut out a very small portion of the beam for these calibrations.

In 1990/1991 NRCC and NIST participated in an absorbed-dose-to-water comparison. A comparison was again performed in 1998. The same NRCC chamber was used in both comparisons with the following NIST/NRCC results:

	1991	1998
Air kerma	0.9941	0.9940
Absorbed dose	1.0036	1.0048

Very little change is seen in the air-kerma calibrations, but the difference in the absorbed-dose comparison could be at least partially due to a difference in field size

Every effort was made to recreate conditions used for Domen's determination of the absorbed dose with the NIST sealed-water calorimeter. Unfortunately, actual field sizes at the measurement distance were not measured at that time, nor was any verification of field size made other than to close the jaws on the jig (*e.g.*, the jaw dial readings were not noted). The only measurement available was a ratio of charges collected with the "set" field size to that with the

It was finally decided that the field size at the time of the water-calorimeter measurements could not be reliably assured. Plotting field size *vs.* calibration coefficients for all NIST absorbed-dose ion chambers, a field size (determined by setting the dials, not using the jig) corresponding to dial settings of 10.75 x 10.75 consistently gave the historical calibration coefficients for the chambers. When the jig was used to set the jaws without any trimmers in the beam, the dials read 10.25 x 10.6. Using these two field sizes, calibration coefficients for all available chambers varied by 0.7 ± 0.04 %.

Because the true field size for the water-calorimeter measurements cannot be independently determined, no corrections appear possible. Rather, assuming that the two settings given above represent the extremes, an additional component of relative uncertainty related to the beam size has been added in the uncertainty assessment.

In summary since the year 2001 a dial setting of 10.75 x 10.75 has been used to define the field size for ion chamber calibrations in terms of absorbed dose to water instead of using the jig that was being used prior to that time.

6.3 Depth Dependence

The source-to-surface distance (SSD) was kept constant as the chamber was moved from a depth of 4 cm to a depth of 6 cm (note that from Fig. 3, SSD = $z - x$). This had a dual effect in that it first altered the distance of the chamber from the source and secondly it adjusted the depth of the chamber in water with each measurement. Figure 12 shows the net current at each position normalized to that seen at the standard 5 cm depth. Due to the precision of the positioning mechanisms, an uncertainty of 0.1 mm would be the maximum uncertainty in the depth measurement. The slope in Fig. 12 shows that this would result in an uncertainty of 0.05 % in the current measurement. One point to note is that absorbed-dose-to-water measurements do not follow the inverse square law. This can be seen in the depth-dose curves generated in characterization of radiotherapy equipment. A least squares fit of the data was used in Fig. 12 to force the data onto a straight line. This is probably not the best fit, as the data actually curve gently in the region shown on the graph. It is felt, however, that the straight-line approximation would be appropriate over the small region in which a correction may need to be made.

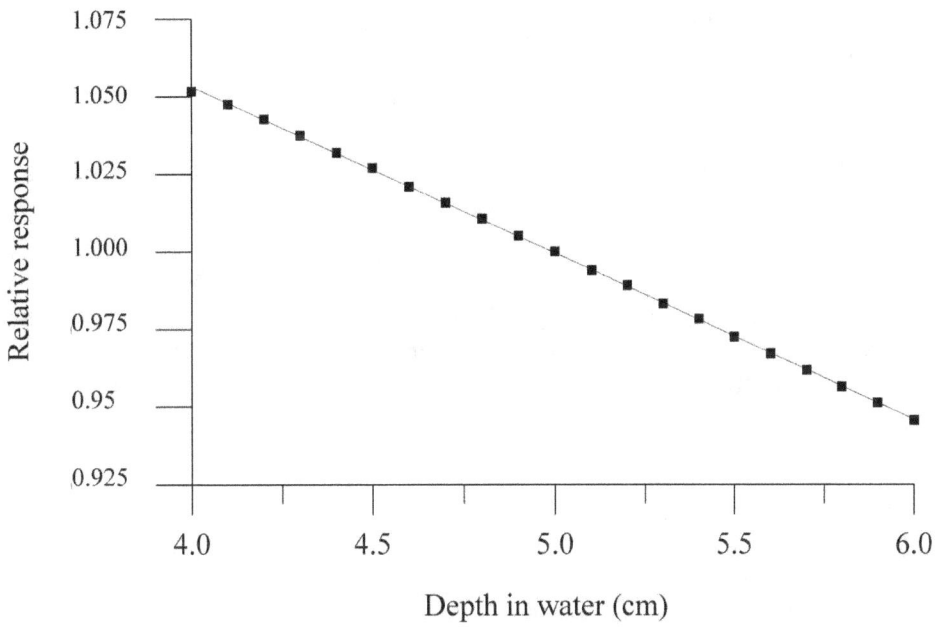

Figure 12. Net Current as a function of chamber depth, normalized to the current at 5 cm depth. Source-to-surface distance (SSD) is maintained at 95 cm.

6.4   Dependence on depth of water above chamber

For this test, the source-to-detector distance (SDD) was maintained at a nominal value of 1 m and the phantom was moved to adjust the depth of water above the chamber, i.e., the surface-to-detector distance. The results of this test can be seen in Fig. 13. As above, the positioning equipment is such that an uncertainty in the determination of water depth would result in an uncertainty in the ionization current measurement of no more than 0.03 %.

6.5   Dependence of chamber distance from bottom of phantom

Using the apparatus mounted in the water phantom, the ion chamber can come no closer to the bottom of the phantom than 9 cm. Water was added for each measurement. To keep the chamber at a constant depth of 5 cm, the chamber and phantom were then moved to maintain the SSD and SDD. The results can be seen in Fig. 14. In essence, there is no effect seen as long as the chamber is at least 10 cm above the bottom of the phantom (no measurements were made at 9 cm). No uncertainty was assigned to this parameter.

6.6   Percentage depth dose

A percent-depth-dose study was performed with the chamber/phantom combination and the results were the same as when the CMS scanning water phantom was used to characterize the source, (see Fig. 10).

6.7   Measurement reproducibility

Reproducibility tests were performed in two ways. First, once the setup was complete, a series of ten calibrations were performed without adjusting anything: a test of the reproducibility of the measurement-chamber system. Second, calibrations were performed over several days and the entire setup procedure was repeated for each calibration, serving as a test of the operator-setup reproducibility. The setup procedure appears to be quite stable with a reproducibility of 0.05 %. Many factors contribute to the setup reproducibility including field size, distance, depth of chamber, and depth of water. The results of this test are shown in Table 2.

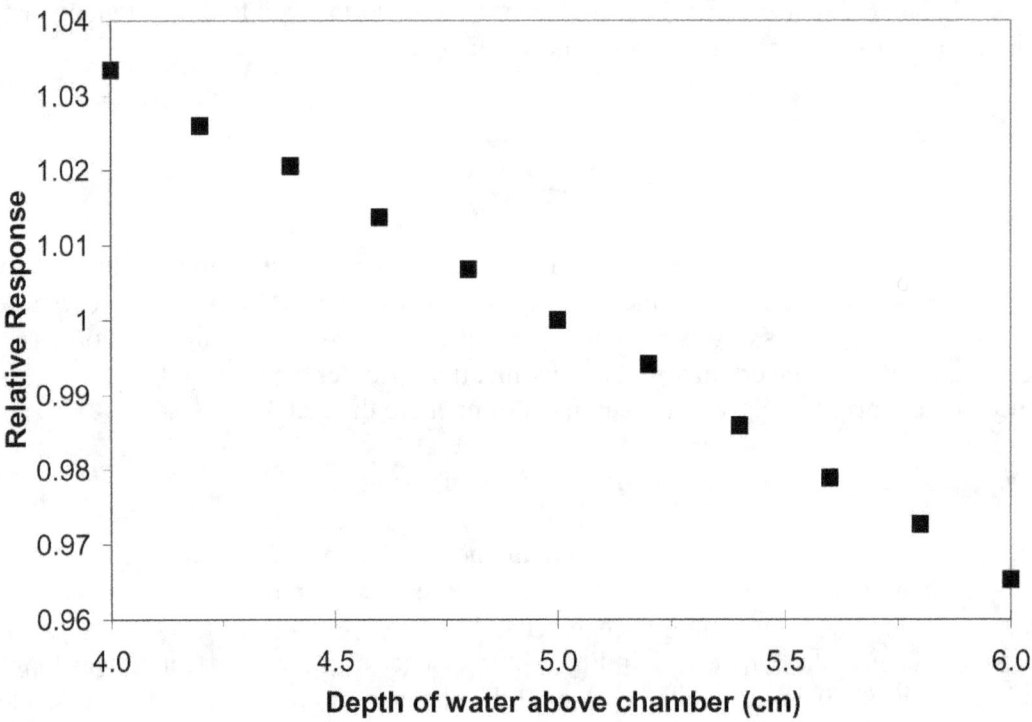

Figure 13. Net current as a function of depth of water above chamber relative to the current with 5 cm water above the chamber. Source-to-detector distance (SDD) is held constant at a nominal 100 cm.

6.8   Electrical leakage measurements

The quality of these chambers coupled with the NIST measurement system results in electrical leakage measurements that do not exceed 0.02 % of the dose readings. Any uncertainty in the leakage would add negligible uncertainty to the total measurement, and therefore an uncertainty for this parameter has not been included.

6.9   Dependence on waterproofing materials

It has been reported[5,10,19,20] that different waterproofing materials affect the response of ionization chambers. Using both chamber types available at NIST, test results using 1 mm acrylic sleeves (for the NE chambers) and rubber/latex sleeves were similar to those reported in the literature. However, only 1 mm acrylic sleeves are used for chambers requiring waterproofing. No uncertainty is applied to this parameter at NIST, but if a customer uses a different waterproofing cap than the one used for the calibration, an additional uncertainty needs to be associated with that change. Using different caps available at NIST (all produced by the same manufacturer), have shown differences in response by up to 0.03 %.

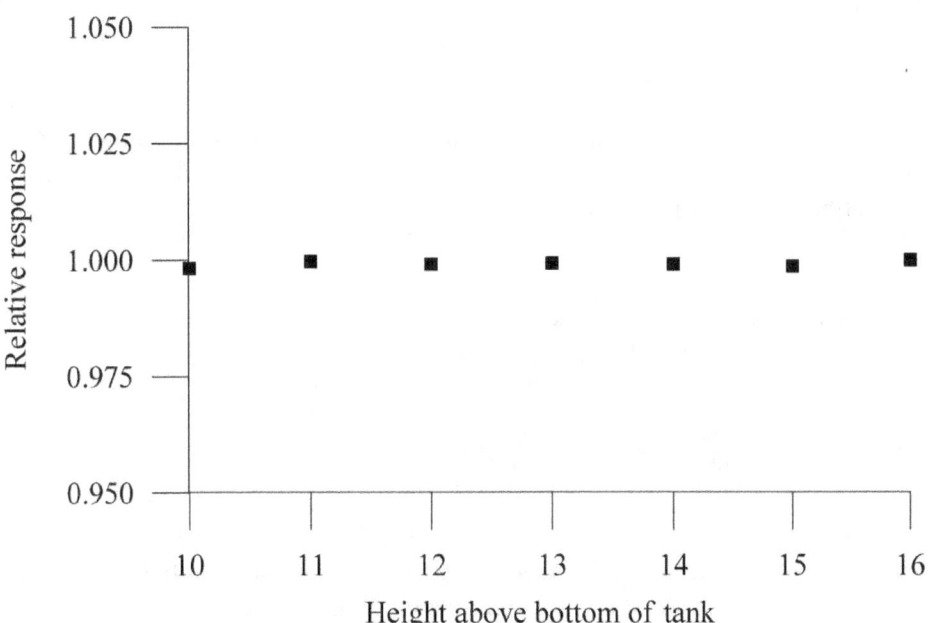

Figure 14. Dependence of measurements on height of chamber above tank bottom. Responses are normalized to that seen at the furthest from the bottom.

6.10   Dependence on water purity and evaporation

Calibrations were performed with both tap water and distilled water filling the phantom. It is recognized that water quality plays an important role in water calorimetry, but it appears to have no effect on the ionization-chamber measurements. Distilled water will be used at NIST because it is readily available and the phantom stays cleaner over time. No uncertainties were assigned for this parameter.

Because of the vertical alignment of the source and chamber, evaporation of water can cause a change in the amount of water above the chamber, see Section 6.4. Water evaporation occurs typically at the rate of 1 mm per 24 hours.

Various "cover" materials, designed to eliminate evaporation, were tested to measure any effect on the charge collected by the ionization chambers. It was impossible to lay materials on the surface evenly or, if one could, they caused some differences in the amount of water over the chambers from that originally measured. The easiest solution appeared to involve laying a piece of Styrofoam over the top of the phantom, as very little difference in charge collection was noticed immediately. However, this caused a layer of saturated air to form between the water surface and the Styrofoam. The density of this air varied with room temperature and increased with time.

All distances are newly set for each calibration so an inherent adjustment is made for evaporation each time. Therefore, no cover is used, and no uncertainty is assigned.

6.11   Polarity effect

The polarity effect was studied and reported in Section 5. Most chambers are calibrated only with negative polarity, unless a special request is made otherwise. Even though there is a difference in response depending on the polarity used, if the chamber is used at the same polarity with which it was calibrated, no correction is necessary. There is no uncertainty associated with this parameter.

At one point, it was thought that the polarity effect might vary with changes in field size. Tests with both types of chambers confirmed that this is not the case.

7.   Operational procedures

NIST accepts chambers that are inherently water proof for calibration in terms of the quantity absorbed dose to water. In the case of chambers that are not waterproof, such as the models NE 2571 and PTW N23333, customers are required to supply an acrylic or nylon sleeve with a maximum thickness of 1 mm. The sleeve must be installed on the chamber prior to being shipped to NIST for calibration.

When a chamber is received at NIST, it is first inspected for any physical damage that may have occurred during shipment. If any damage is detected, the customer is notified immediately, and corrective actions are taken in accordance with the customers' request.

NIST keeps copies of calibration reports for chambers that have been sent for calibration over the years. Therefore a series of past values for the calibration coefficient $N_{D,w}$ may be available and can be used to compare with the new values. Pertinent customer and chamber information is entered into the absorbed-dose-to-water calibration log book.

7.1   Setup

The chamber is mounted as explained previously and biased to a potential appropriate for its operation or as directed by the customer. Typically Farmer-type probes are biased at 300 volts. The chamber is connected to an electrometer used to collect the ionization charge generated during irradiation. The electrometer is part of a data-acquisition system used for this service consisting of a personal computer, two electrometers (model Keithley 617 or Keithley HQ-617), two thermometers and one pressure transducer. This system is used also for the calibration of ion chambers from these same beams in terms of N_k. Details and procedures for the calibration in terms of N_k are described elsewhere[21,22].

The collecting electrode of the ionization chamber to be calibrated is connected via a low-noise cable to the input of either the Keithley 617 electrometer or the Keithley HQ-617 electrometer. Both electrometers are interfaced to the personal computer through an IEEE 488 bus. The appropriate range setting is used to collect data in the data-store mode at 10 second intervals.

Two different temperature probes are connected to the data-acquisition system. A Hart model 1504 readout unit with a model 5611 temperature probe is used, via an IEEE 488 connection, for the absorbed-dose-to-water measurements. A temperature probe supplied as part of a package with the Keithley model 195A multimeter may be used for air-kerma measurements and is also connected by an IEEE 488 bus. A pressure gauge Setra model 370 is used to measure the ambient pressure. The pressure gauge is interfaced to the personal computer through an RS-232 connection.

The software used to control the calibration system has been written in LabView for a personal computer. The Labview software measures the charge generated inside the ion chamber cavity during irradiation for a preset time interval, the water temperature and the ambient pressure. The program provides a value of the ionization current as well as the value of the calibration coefficient $N_{D,w}$. The LabView based software uses Eq. 2 for determining the absorbed dose rate on a daily basis.

7.2 Absorbed-dose-to-water measurements

If the chamber to be calibrated is not inherently waterproof, every effort is made to accommodate the chamber owner's waterproofing sleeve. If the waterproofing sleeve provided will not work in the NIST phantom, the customer is contacted and arrangements are made for the customer to send an appropriate sleeve. In some cases, NIST can use one of its own 1 mm acrylic sleeves on the customer chamber. A latex sheath is then attached to the rear of the NIST sleeve using a heavy rubber band. If this is the case, an additional component in the uncertainty statement must
be included to reflect the use of a sleeve different form the one used by the customer. The chamber is then mounted horizontally in the mounting apparatus and lowered into the water as shown in Figs. 8 and 9.

The temperature probe is mounted in the phantom next to the chamber. As the temperature of the water is generally about 1.5 °C lower than the room temperature, the chamber is allowed time to come to thermal equilibrium. In the case that air-kerma measurements are requested also for the chamber, these measurements usually are done first, prior to the absorbed-dose-to-water measurements, as the ionization chambers come to temperature equilibrium faster when going from the warmer to the cooler environment.

The field size is determined by setting the two pair of collimator jaws to a dial setting of 10.75 x 10.75. A laser located in the pit below the head produces a laser beam aligned with the radiation-beam-center axis. The laser beam is used to align the front edge of the water phantom to the radiation-beam-center axis. Alternatively a plumb bob that hangs from the center of a 50.8 cm x 50.8 cm aluminum block can be used by inserting it in the collimator opening and closing the jaws onto it. Once the front of the phantom is aligned to the radiation-beam-center axis a telemicroscope is used to align the phantom-water surface: Looking through the telemicroscope, one can see a thin black line below a grey area. The cross hairs of the telemicroscope are set on the thin black line that represents the meniscus of the water.

The phantom is then pushed back and an aluminum scale is set in the special holder just below the source head. The platform on which the phantom sits is then raised or lowered until the cross hair of the telemicroscope, and hence the meniscus, is at 53.8 cm on the scale. This setting corresponds to a nominal SSD of 95 cm.

The telemicroscope is then lowered 5 cm to indicate the proper depth for the chamber. The difference between the readings on the telemicroscope is compared to the difference on the scale to ensure that it was lowered by 5 cm. The scale is then removed and the phantom is positioned under the source. Using the laser mounted in the floor, the chamber is centered under the source. The chamber is then raised or lowered by means of the mounting apparatus until it is centered on the cross hairs of the telemicroscope.

The above procedure is performed first with a NIST chamber and then with customer chambers. Because the source is decay corrected rather than using the primary standard to determine the dose rate, the NIST chamber is used to verify setup parameters.

Usually, three leakage measurements, followed by five radiation measurements, and then three more leakage measurements comprise one data set. To determine the net charge, Q_{CAL}, the six leakage measurements are averaged to obtain, $\overline{Q}_L$, and the five radiation measurements are averaged to obtain, $\overline{Q}$. These values are then used as:

$$Q_{CAL} = (\overline{Q} - \overline{Q}_L) \cdot \frac{T + 273.15}{295.15} \cdot \frac{101.325}{P}, \tag{9}$$

where T is the temperature in °C, and P is the pressure in kPa.

Calibration factors for the electrometer, temperature probe and pressure gauge are built into the computer program. The calibration coefficient of the ionization chamber is then calculated as,

$$N_{D,w} = \frac{D}{Q_{CAL}}, \tag{10}$$

where D is the delivered dose.

Absorbed-dose-to-water calibrations are performed with both negative full voltage and negative half voltage. Other measurements can be made at the request of the customer.

7.3 Quality assurance of results

The results of the calibration are compared with previous calibrations when possible. If it is the chamber's first calibration, generally a series of two or three calibrations are performed over a period of a few days to ensure that the chamber is stable. If the calibration coefficient is notably different from the past, a recalibration is performed. The customer is notified immediately if there appears to be a shift in the calibration coefficient, and every effort is made to determine the cause.

Appendix C includes a sample of a typical absorbed-dose-to-water calibration report. The report contains the absorbed-dose-to-water calibration coefficient, $N_{D,w}$, determined with negative full voltage applied to the wall of the chamber. Also listed is the ratio of the absorbed-dose-to-water charges measured with negative full and half voltages applied.

Electronic recombination studies have been performed on NIST chambers, and corresponding correction factors k_{sat} have been determined. For customer chambers, NIST does not include an electronic recombination correction (k_{sat} is set to unity) in the reported value of the calibration coefficient $N_{D,w}$. Each customer has to address the potential impact due to electronic recombination according to the dose rates used at their facilities and their applications.

7.4 Proficiency testing of the ADCLs

Every other year, the AAPM requires the ADCLs to participate in a round-robin proficiency test with the NIST. Following procedures outlined above, NIST calibrates one of its own chambers. The chamber is then sent to each of the three ADCL laboratories for calibration. Once the three laboratories have completed their calibrations, the chamber is returned to NIST where it is calibrated again to ensure that no damage or anomalies occurred while in transit. The two NIST calibration coefficients are averaged for comparison in the proficiency test.

The calibration coefficients reported by the ADCLs are compared to that determined by the NIST. To maintain accreditation by the AAPM, an ADCL must be within specific limits of the NIST value, set by the AAPM.

There have been three ADCL proficiency tests prior to publication of this report. The latest proficiency test involving the ADCLs was conducted in the first quarter of 2004; results of that test are included in a recently published article[23].

8. Uncertainty statement

Table 3 and 4 summarize the uncertainty analysis of the calibration of an ionization chamber in terms of absorbed dose to water.

Table 3 lists the uncertainty components that contribute to the relative combined standard uncertainty of the absorbed-dose rate determined from primary measurements using the NIST water calorimeter. In each case the uncertainties are grouped according to Type A and Type B evaluations.

The value of each of the uncertainty components that appears in Table 3 was determined by Domen in his work published in 1994[5], except for the uncertainty component listed under field size. Note that the value of 0.01 % assigned by Domen for the uncertainty in the thermistor calibration was determined as a result of a detailed study over an extensive period of 184 weeks which is summarized in Fig. 6 of that reference. As a result of that study it was concluded that as long as the thermistors are calibrated before and after each group of measurements done with the water calorimeter, an evaluation of Type B for the uncertainty of 0.01% could be assigned to the calibration of the thermistor.

The additional component listed under field size has been added to account for an uncertainty in the field size resulting from a change in the jaw settings of the ^{60}Co source implemented in the year 2001. The new dial settings of the jaws for absorbed dose-to-water measurements are 10.75 x 10.75. The reason for this change was the discovery of plausible additional non-reproducible collimation that might have existed at the time of the primary measurement in the early 1990s, or been introduced at some time since then, caused by additional beam trimmers that form part of the source.

As described in Section 6.2, the uncertainty for "field size" in Table 3 arises from a problem discovered in the collimator-adjustment procedure for the absorbed-dose-to-water calibrations with the ^{60}Co source located in the low intensity vertical beam facility. In a communication[18] to the 15th meeting of the Consultative Committee for Ionizing Radiation (Section I), an additional component of relative uncertainty of $\frac{0.7\%}{\sqrt{3}} = 0.4\%$ was added in the NIST uncertainty assessment, essentially based on a uniform distribution of uncertainty in the absorbed dose that extends equally above and below the result for our adopted collimator setting by a relative amount of 0.7 %. In retrospect considering a radius of 0.7 % for the uniform distribution seems somewhat too large, since the shift observed is only in one direction. Instead, a more realistic estimate has been made, based on a uniform distribution that adds two contributions: 0.7 % due to the shift observed, plus and additional contribution of 0.1 %, an uncertainty observed in the reproducibility of the 0.7 % result. This results in a radius for the uniform distribution of 0.4 % and therefore in an estimated relative standard uncertainty for the field size of 0.23 % as shown in Table 3.

The relative combined standard uncertainty for the absorbed-dose-to-water rate $\dot{D}$ is obtained by adding in quadrature all the uncertainty components listed in Table 3, resulting in a value of 0.42 %.

Table 4 lists the uncertainty components involved in determining the relative combined standard uncertainty of the calibration coefficient, $N_{D,w}$. Note that this constitutes the uncertainty of the ion chamber calibration. Similar to the above analysis, the value for the relative combined standard uncertainty of $N_{D,w}$ is obtained by adding in quadrature all uncertainty components. The relative expanded uncertainty for the calibration coefficient $N_{D,w}$ is obtained by multiplying the combined standard uncertainty by a coverage factor of 2, resulting in a rounded up value of 1 % as shown in Table 4.

Table 3. Uncertainty analysis for the primary-standard measurement of the absorbed-dose-to-water rate, $\dot{D}_w$. The relative uncertainties are expressed in %.

Uncertainty Components	$\dot{D}_w$	
	Type A	Type B
heat defect		0.30
reproducibility of measurement groups	0.15	
beam attenuation from glass wall		0.10
beam attenuation from calorimeter lid	0.05	
field size		0.23
vessel positioning		0.02
thermistor calibration		0.01
water density		0.02
quadratic sum	0.16	0.39
relative combined standard uncertainty of the absorbed-dose-to-water rate measurement at 5 cm in water	0.42	

Table 4. Uncertainty analysis for the calibration of an ionization chamber in terms of absorbed dose to water. The relative uncertainties are expressed in %.

Uncertainty Components	$N_{D,w}$	
	Type A	Type B
charge	0.10	0.10
time		0.05
air-density correction (temperature and pressure)		0.03
positioning at 1 m reference distance		0.05
positioning at 5 cm reference depth		0.03
measurement reproducibility		0.05
k_{sat}, loss of ionization due to recombination (NIST chambers)	0.01	0.05
humidity		0.06
^{60}Co decay constant[a]		0.05
quadratic sum	0.10	0.17
relative combined standard uncertainty of the chamber reading	0.20	
relative combined standard uncertainty of D_w (from Table 3)	0.42	
relative combined standard uncertainty of the calibration coefficient $N_{D,w}$	0.47	
relative expanded uncertainty of the calibration coefficient, $N_{D,w}$ ($k = 2$)	0.94 ($\rightarrow$ 1.0)	

---

[a]The absorbed-dose-to-water rate determined by the primary-standard instruments has been transferred to the ^{60}Co source and is then decay-corrected to the time of the calibration measurement. For this correction NIST uses a half life of 1925.3 d with a standard uncertainty of 0.5 d.

9. References

1. Seltzer S. M. and Bergstrom P.M., Changes in the U.S. Primary Standards for the Air-Kerma from Gamma-Ray Beams, J. Res. Natl. Inst. Stand. Technol. **108**, 359-381, (2003).

2. Boutillon M. and Perroche-Roux A. M., Re-evaluation of W value for electrons in dry air, Phys. Med. Biol. Vol. **32** (2), pp. 213-219 (1987).

3. Pruitt J. S., Domen S. R. , and Loevinger R., The graphite calorimeter as a standard of absorbed dose for cobalt-60 gamma radiation, J. Res. Natl. Bur. Stand. **86**, 495, (1981).

4. Pruitt J. S., Absorbed-dose calibrations of ionization chambers in a ^{60}Co gamma-ray beam, NBS Special Publication **250-40**, (1990), 50 pp.

5. Domen, S.R., A sealed water calorimeter for measuring absorbed dose, J. Res. Natl. Inst. Stand. Technol. **99**, pp. 121-141, 1994.

6. Shortt, K.R., Report of a comparison of dosimetric measurement standards of the NRC and the NIST, CCEMRI(1)/93-29, (1994).

7. Boutillon, M., Coursey, B.M., Hohlfeld, K., Owne, B., and Rogers, D.W.O., Comparison of primary water absorbed dose standards, Measurement Assurance in Dosimetry, Publication IAEA-SM-330/48, International Atomic Energy Agency, Vienna, p. 95, (1994).

8. P. R. Almond, P. J. Briggs, B. M. Coursey, W. F. Hanson, R. Nath and D. W. O. Roger, AAPM's TG-51 protocol for clinical reference dosimetry of high-energy photon and electron beams, Med. Phys. **26**, pp. 1847-1869 (1997).

9. Chen Mayer H., O'Conor K. W., Minniti R. and Gall K. P., The NIST Room Temperature Water Calorimeter, http://www.arpansa.gov.au/absdos/proc.htm. Proceedings of Workshop on Recent Advances in Absorbed Dose Standards ARPANSA, Melbourne, Australia (2003).

10. Ross, C.K. and Shortt, K.R., The effect of waterproofing sleeves on ionization chamber response, Phys. Med. Biol. **37**, pp. 1403-1411, (1992).

11. Dyk, J.V. and MacDonald, J.C.F., Penetration of high energy electrons in water, Phys. Med. Biol. **17**, pp. 52-65, (1972).

12. Klevenhagen, S.C., Physics and Dosimetry of Therapy Electron Beams, Medical Physics Publishing, Madison, WI, (1993).

13. Aget, H. and Rosenwald, J.C., Polarity effect for various ionization chambers with multiple irradiation conditions in electron beams, Med. Phys. **18**, pp. 67-72, (1991).

14. Tuli, J. K., Nuclear Data Sheets 100, pp. 347 2003; National Nuclear Data Center Brookhaven National Laboratory.

15. Allisy-Roberts, P.J., Boutillon, M., and Lamperti, P.J., Comparison of the standards of air kerma of the NIST and the BIPM for ^{60}Co γ rays, Rapport BIPM-96/9, (1996).

16. Allisy-Roberts, P.J. and Shobe, J., Comparison of the standards of absorbed dose to water of the NIST and the BIPM for ^{60}Co γ rays, Rapport BIPM-98/5, (1998).

17. Shortt, K., Shobe, and J., Domen, S., Comparison of dosimetry calibration services at the NRCC and the NIST, Med. Phys. **27**, pp. 1644-1654, (2000).

18. Shobe, J., Additional Uncertainty in NIST ^{60}Co Absorbed-Dose-to-Water Calibrations, CCRI(I)/01-14, (2001).

19. Hansen, W.F. and Tinoco, J.A.D., Effects of plastic protective caps on the calibration of therpy beams in water, Med. Phys. **12**, pp. 243-248, (1985).

20. Gillin, M.T., Kline, R.W., Niroomand-Rad, A., and Grimm, D.F., The effect of thickness of the waterproofing sheath on the calibraion of photon and electron beams, Med. Phys. **12**, pp. 234-236, (1985).

21. Lamperti P. and O'Brien M., Calibration of X-Ray and Gamma-Ray Measuring Instruments, NIST Special Publication **250-58** (2001), 87pp.

22. Minniti R., Calibration of Radiation Detectors in terms of Air-Kerma using Gamma-Ray Beams, NIST Quality Manual, IRD-P-04 (2004), 25 pp.

23. Minniti R., Chen-Mayer H., Seltzer S., Huq S., Bryson L., Slowey T., Micka J., DeWerd L., Wells N., Hanson W., and Ibbott G., The US Radiation Dosimetry Standards for ^{60}Co Therapy Level Beams, and the Transfer to the AAPM Accredited Dosimetry Calibration Laboratories, Med. Phys. Vol. **33** (4), April 2006.

Appendix A

NIST-BIPM comparison

An indirect comparison of standards of absorbed dose to water of the NIST and of the BIPM was carried out in the ^{60}Co radiation beam at the BIPM in October 1997[16]. The primary standard of the NIST is a water calorimeter and the BIPM primary standard is a graphite-cavity ionization chamber of pancake geometry.

The comparison was undertaken using two NIST ionization chambers as transfer instruments. The result of the comparison is given in terms of the ratio of the calibration coefficients of the transfer chambers determined at the two laboratories. The absorbed-dose-to-water comparison was the first such comparison made directly between the two laboratories.

The conditions of the comparison were defined by the Consultative Committee for Ionizing Radiation (CCRI) and include:
* the distance from source to reference plane at the center of the detector is 1 m,
* the field size in air at the reference plane is 10 cm x 10 cm and the NIST uses 15 cm x 15 cm, the photon fluence rate at the center of each side of the square being 50 % of the photon fluence rate at the center of the square,
* the reference depth in water is 5 gcm^{-2} at the BIPM and 5 cm at the NIST.

The two NIST transfer standards were C552 air-equivalent, conducting-plastic, cavity chambers manufactured by Exradin (Model A12). A collecting voltage of 300 V (negative potential), supplied at each laboratory, was applied to each chamber at least 30 minutes before measurements were made. Chambers were preirradiated for at least 30 minutes before measurements began. The leakage current was less than 0.02 % at both laboratories. During a series of measurements, the water temperature was stable to better than 0.02 °C at the NIST and better than 0.01 °C at the BIPM. The currents were normalized to 295.15 K and 101.325 kPa.

Chamber	NIST $N_{D,w}$ (Gy / µC^{-1})	BIPM $N_{D,w}$ (Gy / µC^{-1})	NIST to BIPM ratio, R	Standard Uncertainty of R
Exradin A12 (1)	49.931	49.981	0.999 0	0.006 0
Exradin A12 (2)	50.122	50.107	1.000 3	0.006 0
		Mean values	0.999 7	0.006 0

Having obtained both N_K and $N_{D,w}$ calibration coefficients for the two transfer chambers, it is interesting to compare these in the beams of the two laboratories. The ratio $N_{D,w} / N_K$ at the BIPM for one of the chambers was 1.1020. At the NIST it was 1.1037, where the chamber was calibrated in air approximately one half meter further from the source than the absorbed dose calibrations. If the chamber had been calibrated one meter from the source for both air kerma and absorbed dose, the ratio would have been on the order of 1.1017. This is based on the average calibration differences seen in all types of chambers between the two locations.

It was concluded that the standards of the BIPM and the NIST are in very good agreement, *(R = 0.999 7, u_c = 0.006 0)*. More information can be obtained from Ref. 16.

Appendix B

NIST - NRCC comparison

In early 1998, three transfer ionization chambers were used to compare the air-kerma and absorbed-dose-to-water calibration coefficients measured by the NRCC and the NIST. Each of the laboratories uses a water calorimeter as the primary standard for absorbed-dose calibrations. The water calorimeter at NRCC is based on the Domen model developed at NIST. Differences between the two are explained in Ref. 17.

Three NE 2571 ionization chambers, supplied by NRCC, were used for this comparison. The results are expressed in terms of the ratio of the calibration coefficients between the two laboratories. A similar comparison was held between the two laboratories in 1991, but the results were never reported. They were, however, included in Ref. 17, and are similar to the results of 1998. This attests to the stability of the absorbed-dose systems at the two laboratories.

The chambers were calibrated in approximately the same manner at the two laboratories. This was verified by Ken Shortt who accompanied the chambers to NIST and participated in the comparison while there. At NIST, the chambers were calibrated under routine calibration conditions with air-kerma measurements made approximately 1.5 m from the source and absorbed-dose measurements 1 m from the source.

The results of the comparison are shown in Table B.1 below. At the time of the comparison, several things were not known by NIST. First, the uncertainty in field size, as discussed in Section 6.2, had not been discovered. Second, the time required for the NE chambers to settle was not known. Though the response shifts over time (see Fig. 9), each data set is fairly tight with a standard deviation of less than 0.02 %. This leads to the false assumption that the chamber has stabilized. Third, the variation in response with distance from the source had not been discovered. These three factors could have had a significant effect on the results. Once all issues have been resolved, further comparisons are in order.

Values for $N_{D,w} / N_K$ are shown in the final column of the table. Further details of this comparison can be found in Ref. 17.

Table B.1

Chamber #	Institute	N_K (cGy/nC)	$N_{D,w}$ (cGy/nC)	$N_{D,w} / N_K$
NE 2571 (1)	NRCC	4.1229	4.5103	1.0940
	NIST	4.0952	4.5331	1.1069
	NRCC / NIST	1.0065	0.9950	
NE 2571 (1)	NRCC	4.1311	4.5186	1.0938
	NIST	4.1069	4.5429	1.1062
	NRCC / NIST	1.0059	0.9947	
NE 2571 (1)	NRCC	4.1741	4.5605	1.0926
	NIST	4.1491	4.5825	1.1045
	NRCC / NIST	1.0060	0.9952	
Average	NRCC / NIST	1.0061	0.9950	

Appendix C

Sample calibration report

The following pages show a sample report only. Portions of the report may be customized to fit a customer's specific need.

National Institute of Standards and Technology

REPORT OF ABSORBED-DOSE-TO-WATER CALIBRATION

FOR

Laboratory Name
Address
Address

Radiation Detection Chamber: Exradin, Model A12, SN ABC

Calibrations performed by [Scientist Name]

Report Reviewed by [Fellow Scientist]

Report approved by [Group Leader]

For the Director
National Institute of Standards and Technology
by

[Division Chief]
Ionizing Radiation Division
Physics Laboratory

Information on technical aspects of this report may be obtained from Ronaldo Minniti, National Institute of Standards and Technology, 100 Bureau Drive Stop 8460, Gaithersburg, MD 20899, (301)975-5586, ronnie.minniti@nist.gov. The results provided herein were obtained under the authority granted by Title 15 United States Code Section 3710a. As such, they are considered confidential and privileged information, and to the extent permitted by law, NIST will protect them from disclosure for a period of five years, pursuant to Title 15 USC 3710a(c)(7)(A) and (7)(B).

Report format revised January 1, 2006

DG:00000-05
SP250 Calibration Service 46110CC. TFN:00000-06
October 3, 2005

National Institute of Standards and Technology

REPORT OF ABSORBED-DOSE-TO-WATER CALIBRATION

FOR

Laboratory Name
Address
Address

Radiation Detection Chamber: Exradin, Model A12, SN ABC

Chamber orientation: The cavity was positioned in the center of the beam with the stem of the chamber perpendicular to the beam direction.
Chamber collection potential: 300 volts (negative charge is collected).
Chamber rotation: The white line faced the source of radiation.
Average leakage: less than [value] % of the radiation measurement current.
Waterproofing Sleeve: not used, as chamber is inherently waterproof.
Environmental conditions: The chamber is assumed to be open to the atmosphere.
Current ratio at full to half collection potential: [value] for an air-kerma rate of [value] Gy/s.
Note on ion recombination: A detailed study of ion recombination was not performed and no correction was applied to the calibration coefficient(s). If the chamber is used to measure absorbed-dose-to-water rates significantly different from those used for the calibration, it may be necessary to correct for recombination loss.

Beam Code	Distance (cm)	Beam Size (cm)	Calibration Coefficient (Gy/C) 295.15 K (22 °C) and 101.325 kPa (1 Atm)	Absorbed-Dose-to-Water Rate (Gy/s)
^{60}Co	100	S15.4	[value]	[value]

DG:00000-05
SP250 Calibration Service 46110CC. TFN:00000-06
October 3, 2005

Explanation of Terms Used in the Calibration Procedures and Tables

Absorbed Dose to Water: The absorbed-dose-to-water rate at the NIST calibration position is measured by a water calorimeter. The ^{60}Co gamma-ray rate is corrected to the date of calibration (from the previously measured value) by a decay correction based on a half-life of 5.27 years.

Waterproofing Sleeve: For the case of non-waterproof chambers, NIST uses a commercially available waterproofing sleeve made of 1 mm PMMA over the collecting volume of the chamber. A latex sleeve is attached to the back of the PMMA sleeve to ensure no water seepage to the chamber.

Absorbed-Dose-to-Water Calibration Coefficient, N_{Dw}: The *absorbed-dose-to-water calibration coefficients*, $N_{D,w}$, given in this report are quotients of the absorbed-dose-to-water and the charge generated by the radiation in the ionization chamber. The average charge used to compute the calibration coefficient is based on measurements with the wall of the ionization chamber at the stated polarity and potential. With the assumption that the chamber is open to the atmosphere, the measurements are normalized to a pressure of one standard atmosphere (101.325 kPa) and a temperature of 295.15 K (22 °C). Use of the chamber at other pressures and temperatures requires normalization of the ion currents to these reference conditions using the normalizing factor F (see below).

Normalizing Factor F: The normalizing factor F is computed from the following expression: $F = (273.15 + T)/(295.15H)$ where T is the temperature in degrees Celsius, and H is the pressure expressed as a fraction of a standard atmosphere. (1 standard atmosphere = 101.325 kilopascals = 1013.25 millibars = 760 millimeters of mercury).

Calibration Distance: The calibration distance is that between the radiation source and the detector center or the reference line. For thin-window chambers with no reference line, the window surface is the plane of reference. The beam size at the stated distance is appropriate for the chamber dimensions.

Beam Size: The beam size is the perpendicular distance from the centerline of the calibration beam to the fifty-percent intensity line. For circular fields, the letter C precedes the dimension; for square fields, the letter S precedes the dimension and the chamber axis is perpendicular to a side of the square.

Half-Value Layer: The ^{60}Co half-value layer (HVL) is 14.9 mm of copper determined from calculations.

Humidity: No correction is made for the effect of water vapor on the instrument being calibrated. It is assumed that both the calibration and the use of that instrument take place in air with a relative humidity between 10 % and 70 %, where the humidity correction is nearly constant.

Uncertainty: The expanded, combined uncertainty of the absorbed-dose-to-water calibration described in this report is 1.0 %. The expanded, combined uncertainty is formed by taking two times the square root of the sum of the squares of the standard deviations of the mean for component uncertainties obtained from replicate determinations, and assumed approximations of standard deviations for all other uncertainty components; it is considered to have the approximate significance of a 95% confidence limit.

Beam Code: The beam code identifies important beam parameters and describes the quality of the radiation field. For gamma radiation, the beam code identifies the radionuclide.

Change in Terminology:

The Radiation Interactions and Dosimetry Group of the NIST Ionizing Radiation Division has made a change in its terminology in calibration and special test reports pertaining to photon and electron dosimetry. This change in terminology is in effect as of 1 May 2002. The proposed changes are based on recommendations in ISO 31-0 (1992) that have been followed for some years now by a number of other international organizations: a quantity with dimensions should be termed a "coefficient," and a quantity that is dimensionless should be termed a "factor."

In this revised terminology, the calibration quantity is defined as the conventional true value of the quantity the instrument is intended to measure, divided by the instrument's reading; this calibration ratio is termed a *coefficient* if it has dimensions or a *factor* if it is dimensionless.

Thus: (a) For our x-ray and gamma-ray calibrations of ionization chambers, for which the calibration ratio has dimensions of gray (or roentgen) per coulomb, the reported quantity is a *calibration coefficient*, rather than the old calibration factor.

(b) For calibrations of instruments that read directly in absorbed dose, kerma or exposure, or their rates, for which the calibration ratio is dimensionless, the reported quantity is a *calibration factor*, rather than the old correction factor.

(c) Other similar calibrations, such as for well-chambers used in brachytherapy dosimetry, will also incorporate these changes.

This change should provide improved clarity in our calibration reports, removing any possible confusion between a reported calibration correction factor (using the old terminology) and those correction factors (*e.g.*, for pressure, temperature, saturation) used in the calibration procedures.

The change in terminology is intended to be benign. *The meaning of the reported calibration quantity has not changed.* The correspondence with the older terminology is outlined above to establish the equivalence of the new terms for those concerned with satisfying, to the letter, documentary standards and protocols.

www.ingramcontent.com/pod-product-compliance
Lightning Source LLC
Chambersburg PA
CBHW081906170526
45167CB00007B/3174